A Bodley Head Young Naturalist

Animals with Poison

GWYNNE VEVERS

Illustrated by Joyce Bee

THE BODLEY HEAD

London · Sydney · Toronto

OTHER TITLES IN THIS SERIES

Animals that Sleep in Winter
Gwynne Vevers

British Library Cataloguing
in Publication Data
Vevers, Gwynne
Animals with poison.—(Young naturalist).
1. Poisonous animals—Juvenile literature
I. Title II. Series III. Bee, Joyce
591.6′9 QL100
ISBN 0-370-30118-8

Printed and bound in Great Britain for
The Bodley Head Ltd
9 Bow Street, London WC2E 7AL
by William Clowes & Sons Ltd, Beccles
First published 1978

Contents

Introduction

Some animals produce poisons in special parts of the body known as glands. These poisons are called venoms.

Animals use venoms to defend themselves, or

attack other animals. Some use venoms to kill prey, that is, other animals killed for food.

Most venoms are injected or forced into the victim by biting or stinging.

Snakes inject venom by biting.

Bees, wasps and scorpions inject venom by a sting.

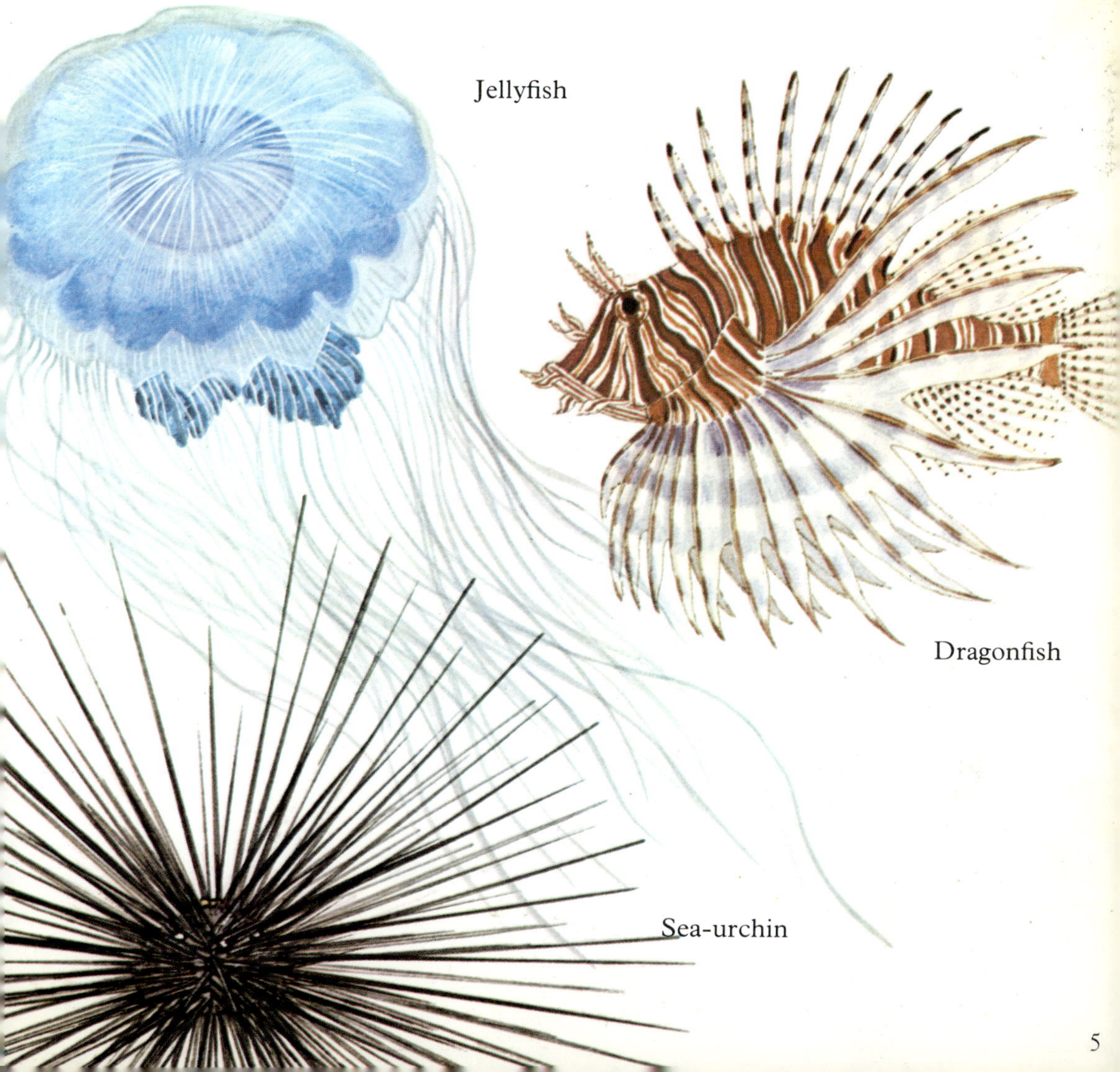

Jellyfish

Dragonfish

Sea-urchin

Snakes

There are almost 3,000 different kinds of snakes. About 400 of these produce venom.

The venom comes from glands in the upper jaw. In most animals these glands produce harmless spit or saliva.

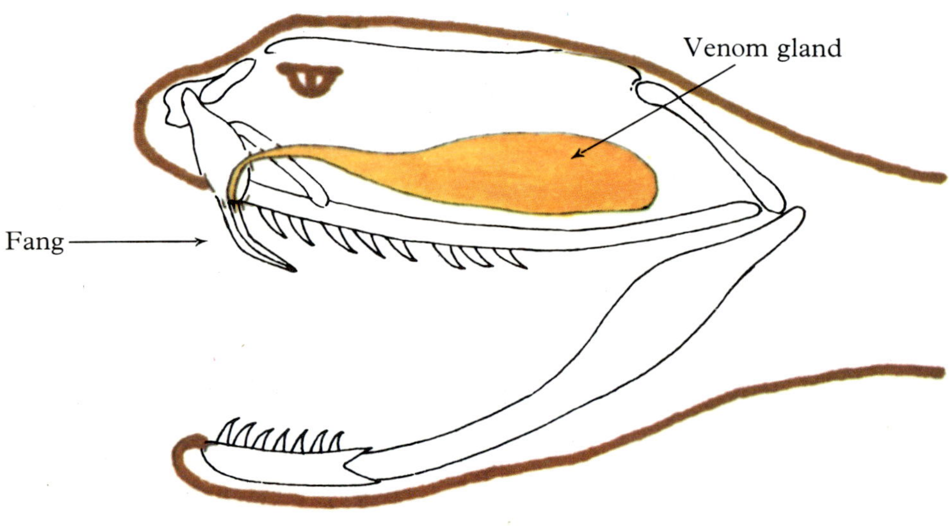

Adder stalking a mouse

In venomous snakes the glands make very powerful venoms. These are used to kill prey for food.

Snake venoms act very quickly. Some venoms paralyse the victim so that it cannot move. Other venoms destroy parts of the victim's body.

The Boomslang

When a snake bites, the venom is squeezed out of the glands by muscles. It passes down a tube or groove in the teeth which are known as fangs.

Some venomous snakes have fangs at the back of the upper jaw.

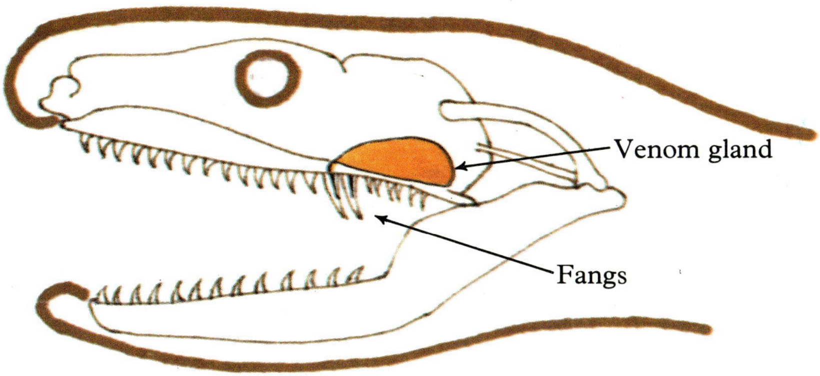

The Boomslang of Africa has very powerful venom and fangs at the back of the jaw. The venom passes down a groove in the fangs and kills the prey.

Cobras, mambas and vipers

Cobras and mambas have fixed fangs at the front of the upper jaw. The venom passes down a canal in the fangs.

Cobras raise their head and spread a 'hood' before striking.

Some cobras spit their venom, often into the eyes of the victim. This may blind the victim.

Cobra spitting venom

Vipers and rattlesnakes also have fangs at the front of the upper jaw. These fangs are hinged and very long. They are folded back when the mouth is shut, but come forwards when the mouth is open.

When approached, a rattlesnake shakes the rattle at the end of its tail as a warning.

Black mamba

Rattlesnake

Lizards

Two kinds of lizard have venom in the lower jaw.

These are the Gila monster from the south-western United States and the Beaded lizard of Mexico. They feed on bird and lizard eggs, or small mammals.

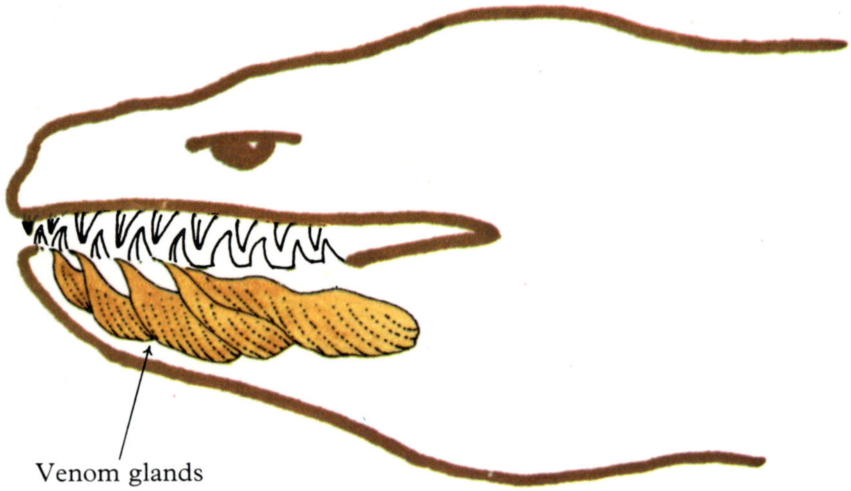

When these lizards bite, the venom flows out between the gums and the lower lips. The lizards really chew the venom into the prey.

Very occasionally the bite can kill a person.

Beaded lizard
Gila monster

Fishes

Sting-rays have a sharp toothed spine on the tail. This spine has a venom gland. When attacked a sting-ray lashes its tail and the spine enters the victim's flesh. The venom flows into the wound, which becomes very painful.

Sting-ray

Dragonfish

Dragonfishes have venom in the spiny fin rays. These can wound an enemy. The venom can kill a human.

Weeverfishes also have venom in their spines. These spines often hurt bathers who step on the fish by accident.

Weeverfish

Fishes and frogs

A few fishes, such as triggerfishes, have poison in their flesh at certain times of the year. This makes them dangerous to eat.

Some pufferfishes have poisonous flesh at all times of the year. They are eaten in Japan, but have to be specially prepared to make them safe to eat.

Triggerfish

Poison Arrow frog

Some frogs and toads have poison in their skin. The Poison Arrow frog has a very powerful poison. South American Indians dip their arrows in this poison. They shoot the arrows into small animals, which become paralysed.

Pufferfish

Other sea animals

Cone snails living on coral reefs have a tooth in the snout that injects venom into prey. The venom acts very quickly.

The sharp spines of some sea-urchins have venom at the tips. The spines protect the sea-urchin. They pierce the skin of an attacker and release their venom, causing great pain.

Jellyfishes and sea-anemones have tiny sting cells in their tentacles. The sting cells, containing venom, are used to catch prey. The sting cells are too small to see.

Some tropical jellyfishes occasionally kill humans.

Cone snail killing butterflyfish

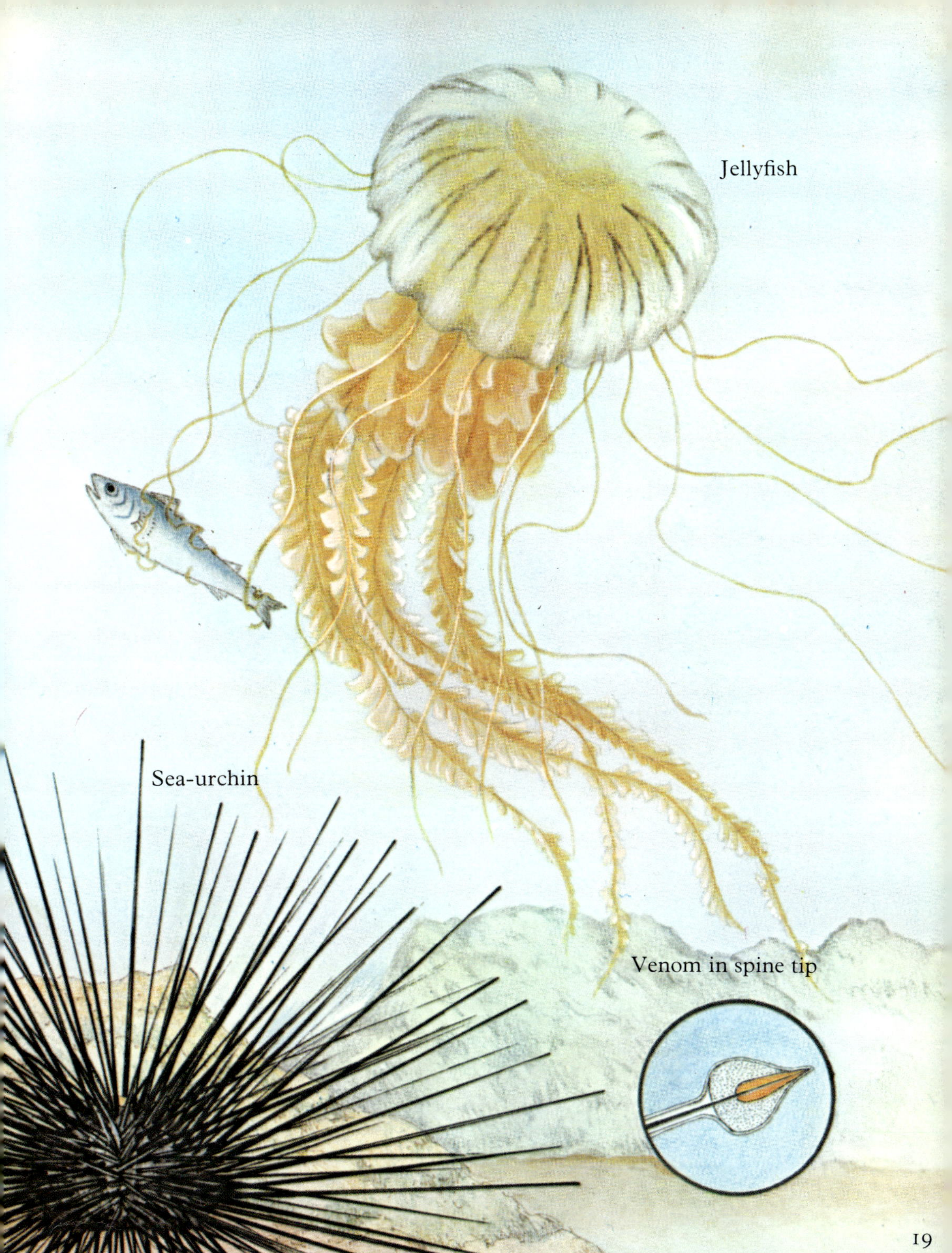
Jellyfish
Sea-urchin
Venom in spine tip

Bees and wasps

Bees and wasps have a sting at the tail end of the body. The sting is joined to a venom gland. The venom is injected when the bee stings.

Bees use the sting to defend their nest.

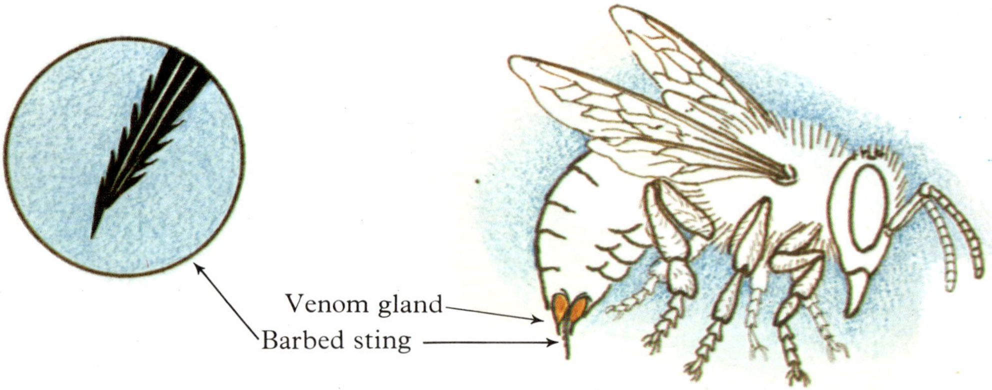

A honey-bee has a barbed sting. When it has stung, the bee flies away, leaving the sting in the victim's skin. The bee later dies.

Honey-bee leaving its barbed sting in a finger

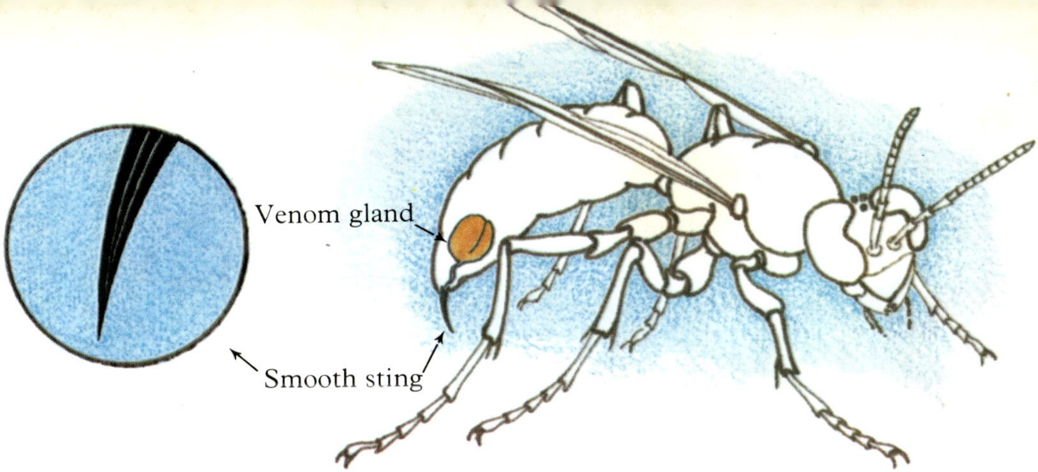

Wasps have a smooth sting. After using the sting a wasp can pull it out and fly away unharmed.

Hunting wasps sting and paralyse a caterpillar or other insect. The wasp then lays an egg in the caterpillar. The egg hatches into a wasp grub which feeds on the caterpillar's body.

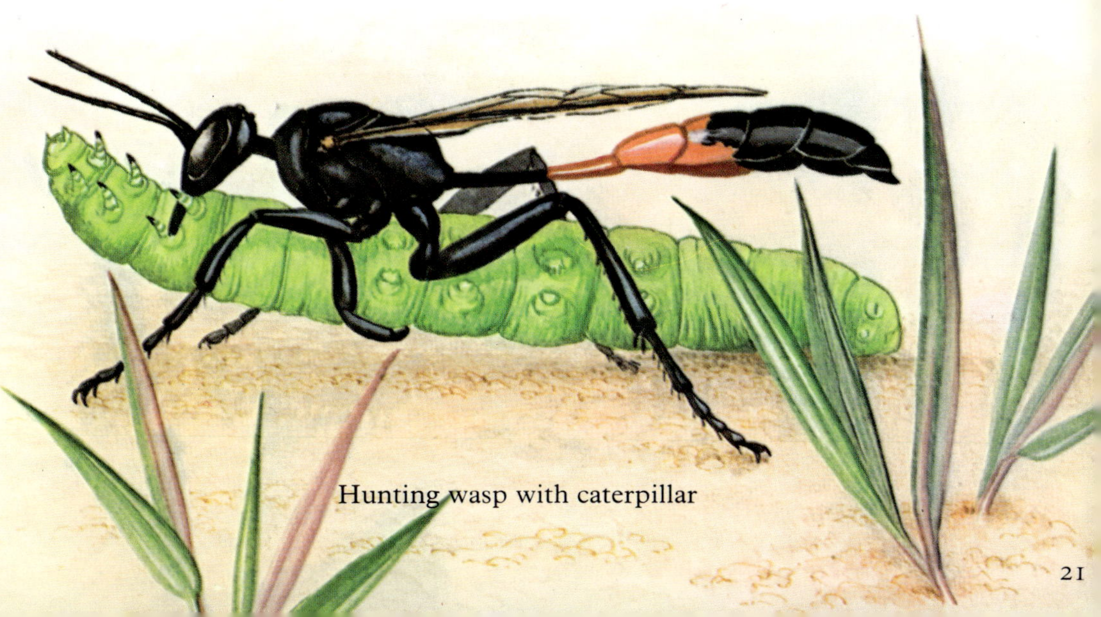

Hunting wasp with caterpillar

Spiders and scorpions

Spiders inject venom into prey with their fangs.

The venom paralyses the prey, usually a fly. The spider often winds threads of silk around the fly, which holds it. The spider then eats the fly.

Some spiders have very powerful venom which can kill a person.

Spider catching a fly

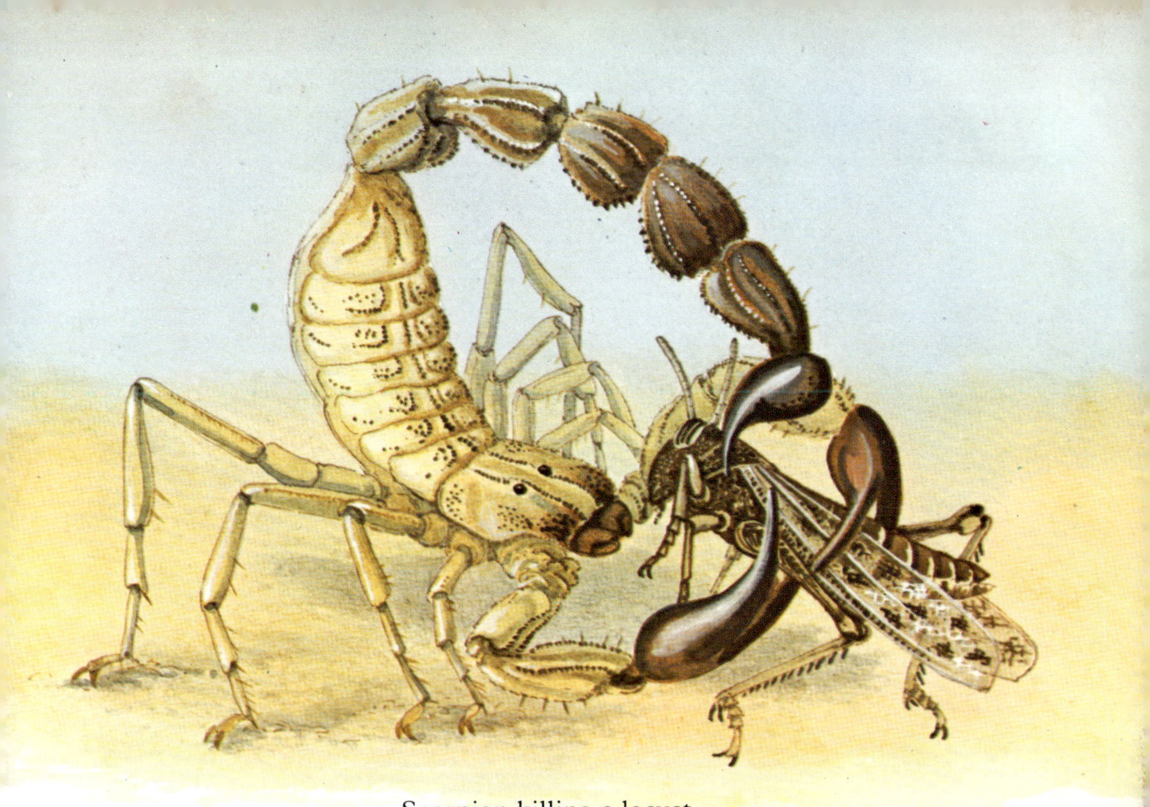

Scorpion killing a locust

Scorpions have a sting at the end of their tail.

They seize prey with their front claws. If the prey struggles the scorpion bends its tail forwards and the sting injects venom into the victim.

A scorpion sting is very painful and sometimes dangerous, or even fatal, to humans.

Index